Management Division 2022

Held at the 2022 AIChE Annual Meeting

Phoenix, Arizona, USA
13-18 November 2022

ISBN: 978-1-7138-7901-5

Printed from e-media with permission by:

Curran Associates, Inc.
57 Morehouse Lane
Red Hook, NY 12571

Some format issues inherent in the e-media version may also appear in this print version.

Printed with permission by Curran Associates, Inc. (2023)

AIChE
120 Wall Street, FL 23
New York, NY 10005-4020

Phone: (800) 242-4363
Fax: (203) 775-5177

www.aiche.org

Additional copies of this publication are available from:

Curran Associates, Inc.
57 Morehouse Lane
Red Hook, NY 12571 USA
Phone: 845-758-0400
Fax: 845-758-2633
Email: curran@proceedings.com
Web: www.proceedings.com

TABLE OF CONTENTS

400b Life and Legacy of a Transformative Nuclear Career

Harold Conner Jr.

Dr. Harold Conner rose from meager beginnings to become the first African-American chemical engineering graduate of his university and, ultimately, leading 3,000 workers at the world's largest uranium enrichment plant. He built a stellar reputation at the helm of a vital national security facility through a challenging period of American history even as he became a guiding beacon and mentor for three generations of nuclear professionals. A product of the Civil Rights era, Dr. Conner is the son of school teachers who instilled in him a lifelong love of learning that helped fuel his ambitious and groundbreaking career pursuits. He broke barriers in academia and in business, leveraging an entry-level internship into a 55-year career in senior management at government nuclear facilities across the country. His story is one of outstanding achievement, culminating in an impressive legacy that reflects well on his professional success and his enduring role as an able ambassador for the field of chemical engineering.

This talk will emphasize the Leadership Traits that have been developed over my career of Challenging the Status Quo, inspiring a Shared Vision, Enabling Others to Act, Modeling the Way, and Encouraging the Heart. These leadership principles have ensured my career success and can help you succeed in any professional endeavor.

400c Overcoming Challenges in First-Ever Demolition of a Gaseous Diffusion Complex

Kenneth Rueter

In this presentation, we will hear how UCOR, an Amentum-led partnership with Jacobs, solved a major challenge for the U.S. Department of Energy by completing the complex and formidable task of the first-ever decommissioning, deactivation, demolition and remediation of the world's first uranium gaseous diffusion plant—the K-25 site in Oak Ridge, Tennessee, now known as the East Tennessee Technology Park (ETTP).

UCOR President and CEO Ken Rueter will describe how the UCOR workforce overcame a challenging work environment caused by structural degradation, radioactive contaminants, hazardous chemicals, materials and wastes, biological hazards, wildlife, insects and environmental conditions. UCOR was able to complete the mission—called Vision 2020—four years ahead of DOE's schedule and $80 million under budget, resulting in a $500 million savings to taxpayers.

Through environmental cleanup of ETTP, UCOR is facilitating DOE's vision of transforming once-contaminated government land into a multi-use industrial park that will boost economic growth. The new, revitalized site is home to new industry, historic landmarks—including a new national park—and conservation areas that honor the past while supporting development.

400d First Ever Nuclear Decommissioning Minor in U.S.

Wesley Hines

The first nuclear decommissioning and environmental management minor offered by any major educational institution in the U.S. is drawing student interest and industry praise as the number of commercial nuclear power plants and government defense facilities slated for shut down and cleanup continues to grow. The focus on nuclear decommissioning is the result of a collaboration among the University of Tennessee, Oak Ridge Associated Universities and UCOR, an Amentum-led partnership with Jacobs and lead Department of Energy contractor for cleanup of the Oak Ridge Reservation in Tennessee.

Developing and maintaining a trained workforce to work in hazardous nuclear environmental cleanup projects is essential to the future. The cleanup industry needs the leaders that the UT program is producing. In Oak Ridge alone, there are jobs for two-plus generations of workers in environmental cleanup. The DOE annual cleanup budget currently is approaching $8 billion. Add to this research and test reactors, uranium recovery sites and fuel cycle facilities and Department of Defense sites, and the number grows by billions more and extends over many decades.

400e Nuclear Facilities Operations Expertise at the Savannah River Site and the Lawrence Livermore Laboratory

Howard Walls Sr.

A nuclear engineer with an Executive MBA from Stanford University, Howard Walls will discuss how Dr. Harold Conner served as a mentor when it came to managing and leading the complex Nuclear Facilities Operations at key Nuclear Facilities in the Department of Energy.

Walls will recount his nearly four decades of managerial experience in Nuclear Facilities Operations. He formerly served as vice president and director of Strategic Planning and Mission Integration for Westinghouse Savannah River Company, which operated the Savannah River Site in South Carolina for the U.S. Department of Energy. He was responsible for directing the development of strategies in support of the Savannah River Site's present missions and the integration of new missions. He was named Deputy to Dr. Conner who ran all Nuclear Operations except for the Canyons at the Savannah River Site.

Walls was recruited by Dr. Conner for the contract transition for URS, Bandamp;W, Bechtel, and Texas Aandamp;M to manage and operate the Lawrence Livermore National Laboratory in Livermore, California. Conner was named Associate Director for the Facilities and Infrastructure Directorate, Howard Walls served initially as Facilities and Infrastructure Directorate Maintenance Manager for all Nuclear and non-nuclear Facilities. Ultimately Mr. Walls was named the Deputy Associate Director of the Facilities and Infrastructure Directorate, reporting to Dr. Conner.

Howard Walls, as the Senior Vice President of Business Development of Westinghouse Government Services Business Unit, was instrumental in recruiting Dr. Conner away from the Oak Ridge Operations of the Department of Energy to join the Savannah River Site Operations and will relay that story.

Walls holds Bachelor of Science degrees in physics and in mathematics from Tougaloo College and a Master of Science degree in nuclear engineering from Tuskegee Institute.

127a Establish Psychological Safety to Enhance Team Creativity, Inclusivity and Retention

Dennis Hess

Effective and efficient technical team operation depends upon positive interactions among team members so that the diverse abilities and experiences represented can be used to advantage while ensuring the engagement and motivation of each member. Challenging problems and a clear understanding of each member's role in the effort facilitate realization of these goals. However, maximizing individual contributions requires psychological safety within the team. Psychological safety refers to the feeling that any team member can challenge the status quo, question leadership directions, take risks, and offer novel ideas without derision or retribution. Studies have demonstrated that when psychological safety exists, team performance improves, employee satisfaction is enhanced, innovation and creativity increase, effective learning from mistakes occurs, and inclusivity is supported. Leaders set the tone and promote this culture through their actions, policies and character. This talk will describe what psychological safety is and is not, what leaders can do to ensure that this characteristic is pervasive within their team/organization, and why benefits result.

199a Impact of Environmental Regulatory Compliance: Corporate Responsibility and Economic Consequences

Rengasamy Kasinathan, Madalyn Bozinski

Process industries, like refineries, operating many sources such as reactors, scrubbers, boilers, cogen units, etc. must comply with the requirements of various regulatory agencies having jurisdiction (AHJ). Although the processes and associated equipment are operated in accordance with design and operating conditions, strict compliance with applicable codes as stipulated in their operating permit conditions specific to the equipment are equally important. Published data suggest that many industries have not given importance to compliance, because of the fact that they have to meet production targets and maximize profit. More often than not, corporate management is not accustomed to the impact of noncompliance unless they are audited unannounced, and hefty penalties are levied. Many corporate managers mistakenly think that in the post-COVID era, compliance is not a priority.

The US Environmental Protection Agency (EPA) has authority from US Congress by which a violation can be fined up to $37,500 a day, up to 180 days, depending upon the severity of the violation. In addition, local AHJs can impose fines, too. In some cases, litigation for causing adverse environmental and human health and safety is also possible. Despite mitigating the violations, keeping the facility continually in compliance is a challenge due to the required additional resources and associated costs. Some industries use site specific compliance tracking systems. However, they fail to keep up with the growing regulations that affect their operations. Therefore, facilities must take regulatory compliance very seriously, not only to avoid fines, but to meet their economic goal and not be exposed to the public as a "big polluter, with no corporate social responsibility".

This paper assesses the compliance status of a few industries based on open source data collected from the AHJs, types of violations issued, mitigation efforts taken, and areas of future potential violations. Recommendations are made as to how such facilities can remain in compliance, responsibilities of the operating personnel, and audit methods including periodic inspections. A specific compliance tracking software tool is discussed to aid the facilities to stay in compliance.

199b Chemical Engineers in Academia.

Stephen P. Beaudoin

Stephen Beaudoin is Director of Purdue Energetics Research Center (PERC) and Professor at the Davidson School of Chemical Engineering at Purdue University. He founded PERC and established it to be one of the largest and most impactful academic center of its kind in the western hemisphere, with a commanding international reputation. Professor Beaudoin has authored 109 refereed publications covering the areas of particle, powder, and thin film adhesion, with applications in explosives detection and optimization, and in food, pharmaceutical and microelectronics manufacturing. He is also a multi-awarded educator and mentor. He obtained his degree in B.S. in Chemical Engineering from the Massachusetts Institute of Technology, M.S. in Chemical Engineering from the University of Texas Austin, and Ph.D. in Chemical Engineering from North Carolina State University.

199c Perspectives on Leadership

Levi Thompson

Levi Thompson is Dean of the College of Engineering and Elizabeth Inez Kelley Professor of Chemical Engineering at the University of Delaware where he leads 3,600 students, 300+ faculty and staff, seven departments and more than a dozen research centers in addressing some of the world's most vexing challenges.Prof. Thompson's scholarly research on nanostructured materials for catalytic and energy storage applications is described in 150+ publications and more than 10 patents describing his research. He co-founded two companies. Prof. Thompson is a member of the National Academy of Engineering, an AIChE Fellow and recipient of numerous awards his research, entrepreneurship, and teaching. He earned a B.Ch.E. from the University of Delaware as well as M.S.E. degrees in Chemical Engineering and Nuclear Engineering and a Ph.D. in Chemical Engineering from the University of Michigan.

323c Building the Future Sustainable Workforce

John F. Cirucci

Since the industrial revolution, engineering and applied sciences have evolved to fulfill changing human needs. Societal stewardship of our atmosphere now demands significant shifts in industrial technologies and practices. This requires a future sustainable workforce equipped with skills and knowledge to address these needs. The rate of change can no longer be simply addressed through workforce attrition and replacement but rather through new models for continuous learning, adaptation and collaboration.

AUTHOR INDEX

NOTES

NOTES